CW00393461

THE CLIMATE OF SCOTLAND
SOME FACTS AND FIGURES

For enquiries regarding commercial forecast services in Scotland, contact:

Glasgow Weather Centre
33 Bothwell Street
GLASGOW
G2 6TS
Telephone: 041-248 7272

Or

Aberdeen Weather Centre
Seaforth Centre
Lime Street
ABERDEEN
AB2 1BJ
Telephone: Aberdeen (0224) 210571

The Meteorological Office

THE CLIMATE OF
SCOTLAND
SOME FACTS AND FIGURES

London: Her Majesty's Stationery Office

Sunset over Eigg and Rhum

SHETLAND

• Lerwick

Climatological stations in Scotland
for which data are given
in this publication.

ORKNEY

• Kirkwall

WESTERN
ISLES

• Wick

• Stornoway

• Ullapool

HIGHLAND

Benbecula

Prabost

Achmore

Nairn

• Gordon Castle

GRAMPIAN

Portree

Glenmore Lodge

• Craibstone

Plockton

Cairngorm

Fort Augustus

Braemar

Onich

• Pitlochry

• Arbroath

TAYSIDE

Tiree

Perth

St. Andrews

Aros

Callander

FIFE

Inverary Castle

CENTRAL

Edinburgh

• Dunbar

Abbotsinch

LOTHIAN

Eallabus (Islay)

(Glasgow Airport)

Penicuik

• Floors Castle

STRATHCLYDE

BORDERS

Prestwick Airport

Auchincruive

• Eskdalemuir

DUMFRIES

Loch Dee

&

• Dumfries

GALLOWAY

West Freugh

THE CLIMATE OF
SCOTLAND
SOME FACTS AND FIGURES

Photograph courtesy of Scottish Tourist Board

Cover photographs courtesy of:
(Top left) Marjory Roy
(Top right) Marjory Roy
(Bottom left) Alex Gillespie
(Bottom right) Scottish Tourist Board

Looking south from Beinn Alligin on a clear day

Introduction

This brief description of Scotland's climate has been compiled with a number of interests in mind. These include the tourist industry, school pupils, and organizations and businesses which are considering Scotland as a location for their operations. Much more detailed information and analyses for particular areas or locations may be obtained from publications in the 'Climate of Great Britain' series or from:

The Superintendent
Meteorological Office
Saughton House
Broomhouse Drive
EDINBURGH
EH11 3XQ
Telephone: 031-244 8362/8363

**Requests for climatological information
for England and Wales
should be addressed to:**

The Director-General
Meteorological Office (Met O 3b)
London Road
BRACKNELL
Berkshire
RG12 2SZ
Telephone: Bracknell (0344) 420242 Ext 6207

and for Northern Ireland to:

Senior Meteorological Officer
Belfast Climate Office
1 College Square East
BELFAST
BT1 6BQ
Telephone: Belfast (0232) 228457

In addition the Met. Office provides a wide range of forecast services to the public, industry and commerce through a network of Weather Centres. Details can be obtained from:

The Met. Office
Met O /
London Road
Bracknell
Berks
RG12 2SZ
Telephone: Bracknell (0344) 854498

Photograph courtesy of Mike Porter

Rainfall

There are a number of aspects of Scotland's climate which are not generally known or understood, but perhaps the main misconception is that the whole of the country suffers from very high rainfall. In fact long-term averages of rainfall calculated from actual measurements (**Figure 2**) show that over 6000 square kilometres of Scotland the annual rainfall is less than 800 mm (31 inches). Many districts in the north and east of Scotland have, on average over the four summer months of May, June, July and August, a total rainfall of less than 250 mm (10 inches) — an average which compares closely with the total rainfall for these same months in the drier parts of England.

Monthly and annual averages of rainfall for the period 1951–80 for a selection of places in Scotland are given in **Table 1**. For comparison, annual averages for a number of places in England, Wales and Northern Ireland are shown in **Appendix 1**. From these figures it can be seen that each area in England and Wales can be matched by one in Scotland, e.g. London by Edinburgh, the English Lake District and the Welsh Mountains by the Scottish Highlands, the region around The Wash by the region around the Moray Firth and so on.

It must be admitted that the ruggedly scenic areas of Scotland are very wet, and because the high hills cover a larger area than they do in England and Wales there is a larger area in Scotland with an annual rainfall exceeding 1600 mm (63 inches). But it is the heavy rainfall in these areas which provides the natural resource for the generation of hydro-electricity, and creates some of the most attractive features of the landscape with its fast flowing rivers and spectacular waterfalls.

One fact that is not widely appreciated, however, is the very marked seasonal variation in average monthly rainfall in the west of Scotland, with the total rainfall for the five months from February to June being only about 55% to 60% of that from September to January. The seasonal difference is less marked in the east of Scotland (where on average July and August are the wettest months in the year). This is illustrated by **Figure 1** which shows the average rainfall for each month at Tiree in the Hebrides and at Edinburgh (Royal Botanic Garden). Thus a regular visitor to the west of Scotland would have a much greater chance of enjoying dry, settled weather if he chose to come in the spring and early summer.

It can be seen from **Figure 2** that the low-lying strips of land around the coast of Scotland are relatively dry and these often escape the heavy and prolonged showers which occur over high ground. From the point of view of scenery, especially when seen from a car, the most dramatic effects are often found in showery weather when visibility between the showers is usually excellent.

Figure 1. Average monthly rainfall (1951–80)

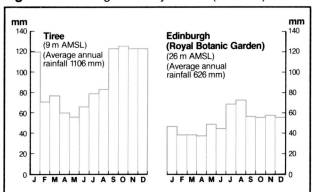

River in spate at Hermitage after heavy rain

4

Thunder

The frequency of summer thunderstorms and hail in Scotland is low, and considerably less than that in much of England. For example, on average, Edinburgh and Glasgow have about 7 days per year with thunder while many places in England have an annual average of 15 to 20 days. Along the coastline of the Moray Firth the average is only 3 or 4 days per year.

Photograph courtesy of Scottish Tourist Board

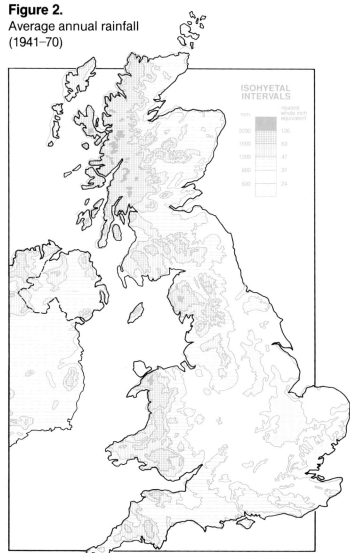

Figure 2.
Average annual rainfall
(1941–70)

ISOHYETAL
INTERVALS

mm	nearest whole inch equivalent
3200	126
1600	63
1200	47
800	31
600	24

5

Table 1. Monthly and annual averages (1951–80) of rainfall for a selection of stations in Scotland

Location	Altitude (metres)	Jan	Feb	Mar	Apr	May	Jun	Jul	Aug	Sep	Oct	Nov	Dec	Year
Shetland														
Lerwick	82	127	93	93	72	64	64	67	78	113	119	140	147	1177
Orkney														
Kirkwall	26	105	71	72	54	53	57	55	77	88	110	120	121	983
Western Isles														
Benbecula	5	129	86	89	62	65	76	83	83	119	139	140	132	1203
Stornoway	15	115	77	80	66	62	67	72	74	103	126	129	125	1096
Highland														
Ullapool	12	119	84	93	76	73	83	80	92	115	155	159	161	1290
Achmore (nr Plockton)	18	104	68	70	56	56	63	78	82	107	112	124	112	1032
Portree (Skye)	27	182	116	129	93	91	104	113	118	170	204	203	210	1732
Onich (nr Fort William)	15	200	132	152	111	103	124	137	150	199	215	220	238	1981
Wick	36	81	58	55	45	47	49	61	74	68	73	90	82	783
Fort Augustus	21	112	72	83	58	72	61	65	82	103	124	126	140	1098
Nairn	8	48	34	33	36	43	46	62	75	50	54	60	52	593
Glenmore Lodge	341	96	64	71	68	76	73	87	107	80	96	100	106	1024
Grampian														
Gordon Castle	32	55	44	42	42	51	55	77	89	57	66	75	67	720
Braemar	339	93	59	59	51	65	55	58	76	73	87	87	96	859
Craibstone (nr Aberdeen)	102	77	57	54	51	62	54	79	83	66	78	80	80	821
Tayside														
Pitlochry	144	94	59	52	47	64	62	65	72	70	76	69	94	824
Arbroath	29	50	40	40	38	48	44	62	67	50	52	54	54	599
Perth	23	70	52	47	43	57	51	67	72	63	65	69	82	738
Fife														
St. Andrews	18	62	48	44	39	50	44	61	68	55	55	63	64	653
Lothian														
Edinburgh (Royal Botanic Garden)	26	47	39	39	38	49	45	69	73	57	56	58	56	626
Dunbar	23	46	33	33	33	47	41	55	68	47	48	55	49	555
Borders														
Floors Castle (Kelso)	59	55	43	38	40	55	51	63	75	57	59	63	57	656
Central														
Callander	107	164	113	115	81	97	86	97	113	141	147	175	193	1522
Strathclyde														
Tiree	9	120	71	77	60	56	66	79	83	123	125	123	123	1106
Inveraray Castle	12	210	132	150	115	106	124	137	148	209	225	228	252	2036
Eallabus (Islay)	21	139	85	92	70	67	74	93	93	128	139	149	149	1278
Glasgow Airport	5	96	63	65	50	62	58	68	83	95	98	105	108	951
Auchincruive (nr Ayr)	45	82	50	53	47	51	58	80	91	100	97	99	93	901
Aros (Mull)	37	210	116	142	97	96	109	120	133	199	208	203	220	1853
Dumfries and Galloway														
West Freugh (nr Stranraer)	15	100	65	67	53	55	57	70	82	98	100	105	103	955
Loch Dee	244	236	154	168	130	121	122	142	184	214	222	241	263	2197
Dumfries	49	103	72	66	55	71	63	77	93	104	106	109	104	1023
Eskdalemuir	242	150	99	108	86	94	95	107	122	142	140	153	160	1456

Rainfall Conversion: 1 inch = 25.4 mm

Millimetres	Inches	Millimetres	Inches	Millimetres	Inches
0	0.0	40	1.6	600	23.6
5	0.2	45	1.8	700	27.6
10	0.4	50	2.0	800	31.5
15	0.6	100	3.9	900	35.4
20	0.8	200	7.9	1000	39.4
25	1.0	300	11.8	1500	59.1
30	1.2	400	15.7	2000	78.7
35	1.4	500	19.7	4000	157.5

6

Visibility

Scotland enjoys remarkably good visibility since the greater part of the country is remote from the industrial and populous areas of Britain and continental Europe. Smoke fogs are now rare even in the industrial areas of Central Scotland, where the decline in traditional heavy industry and the move away from open fires for domestic heating have greatly reduced the local pollution. Water-droplet fogs may form overnight in low lying inland areas on calm, clear nights, particularly in winter, but generally these clear during the day.

When poor visibility occurs at places on or near the east coast of Scotland or in the Northern Isles the cause is, more often than not, a sea fog from the North Sea, which is known locally as *haar*. Haar occurs from time to time from April to September and can ruin what would otherwise be a brilliantly fine day. Fortunately the haar does not usually penetrate very far from the coast (though it can reach Glasgow through the Central Lowlands, and Fort William down the Great Glen) and it tends to break up inland during the day. Indeed the presence of haar along the east coast usually signals the presence of brilliantly fine weather in central and western areas of Scotland.

Figure 3 shows for Tiree, Edinburgh Airport and London (Heathrow) Airport the relative frequency of visibility in different ranges at 12 GMT for each month of the year during the period 1957–76. (The observed visibility is the *lowest* seen in any direction from the observing point.) At that time of the day early morning mist or fog has normally cleared.

At Tiree not only is the visibility generally very good (as might have been expected) but there is very little variation in the frequency pattern throughout the year. By contrast both Edinburgh Airport and London Airport show a marked seasonal variation with a much greater frequency in the winter of visibilities of 10 km or less. The small number of occasions at Edinburgh with visibilities of 1000 m or less in the summer are likely to have been due to persistent haar (there were none in summer at London Airport). Edinburgh has a greater frequency of good visibility (more than 30 km) than London in each month of the year.

In certain weather situations (for example with moist south-westerlies covering the country) the cloud base can be quite low, especially in the west, with all high ground being shrouded in cloud. The resulting hill fog can give very low visibilities which are not only a potential hazard for hill-walkers, but also for motorists since many of Scotland's roads have considerable changes in height which may take them in and out of hill fog.

Haar (sea fog) breaking up to the lee of high ground in Shetland

Photograph courtesy of Alan Gair

Figure 3. Distribution of visibility observations

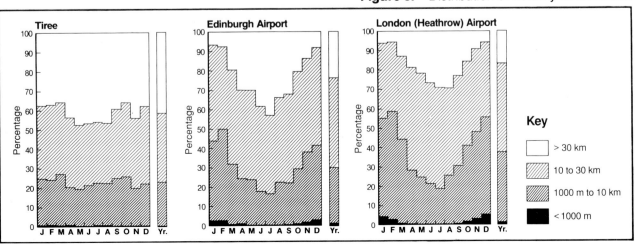

Key

☐ > 30 km
▨ 10 to 30 km
▧ 1000 m to 10 km
■ < 1000 m

7

Snow

The development of skiing facilities in Scotland and quite frequent references on television, radio and in the Press to certain roads being blocked by snow have given rise to a common misconception that the whole of Scotland suffers from very severe, snowy winters. Monthly and annual values for the average number of days with snow lying at 09 GMT for a selection of stations (**Table 2**) show that this is far from the case. On low ground in the Western Isles and in most coastal areas of Scotland snow lies on average for less than 10 days in the year, although this goes up to between 15 and 20 days in the north and north-east. In England and Wales the values are around 1 to 2 days in the south-west and around 5 to 10 days elsewhere. Serious difficulties with roads blocked by snow are not common on *low ground* anywhere in the United Kingdom, though there are considerable variations from year to year. For example at Glasgow Airport there were no days with snow lying in the winter of 1960/61, but 47 in 1946/47.

Skiing in late April on long-lasting snowbeds on the Braeriach plateau

Photograph courtesy of Marjory Roy

As will be explained later, temperature generally falls with height, and precipitation which reaches the ground at low-level sites as rain may fall as snow over high ground. As a result there is a marked increase with height in the number of days with snow falling and the number of days with snow lying. On the A74/M74 road which is the major route between Central Scotland and Carlisle in England the average number of days with snow lying at the highest point (Beattock Summit, 315 m) is about 38 days, and at Drumochter Pass (460 m) on the A9 north to Inverness it is about 70 days. The roads which are most commonly reported as being blocked by snow are the A93 from Blairgowrie to Braemar, which reaches a height of 680 m at The Cairnwell, and the A939 between Cockbridge and Tomintoul which reaches 635 m at the Lecht.

One of the features of Scotland's climate is the windiness of the winter months and this has a major effect on the pattern of snow cover. When snowfall is accompanied by strong winds the snow is deposited on leeward slopes and in the lee of obstructions such as walls, buildings, woods and snow-fencing, with exposed areas often being left bare. Natural hollows become filled to a considerable depth and it is the existence of suitable high-level corries which has led to the development of a skiing industry in Scotland. It is rare for a complete snow cover to persist for a considerable period of time except near the summits of the highest mountains (on average snow lies for about 6½ months on the tops of Ben Nevis and Ben Macdui), but snow beds are much more persistent and many survive well into the summer. Some of these are semi-permanent and only disappear in very occasional summers.

Table 2. 30-year (1951–80) average number of days in month with snow lying at 0900 hours at selected Scottish stations

Location	Altitude (metres)	Jan	Feb	Mar	Apr	May	Jun	Jul	Aug	Sep	Oct	Nov	Dec	Year
Shetland														
Lerwick	82	7.5	7.8	3.8	1.4	0.1	0.0	0.0	0.0	0.0	0.2	2.4	4.9	28.1
Western Isles														
Stornoway	15	3.5	2.8	1.4	0.2	0.0	0.0	0.0	0.0	0.0	0.0	1.0	1.5	10.4
Highland														
Wick	36	5.4	5.6	2.2	0.5	0.0	0.0	0.0	0.0	0.0	0.0*	1.5	3.0	18.2
Nairn	8	5.5	5.6	1.7	0.2	0.0	0.0	0.0	0.0	0.0	0.0	0.8	2.7	16.5
Grampian														
Braemar	339	15.6	15.9	9.7	2.4	0.1	0.0	0.0	0.0	0.0	0.3	4.6	10.4	59.0
Craibstone (nr Aberdeen)	102	9.6	9.6	4.9	0.9	0.1	0.0	0.0	0.0	0.0	0.2	2.6	4.7	32.6
Tayside														
Perth	23	5.7	5.2	1.2	0.1	0.0	0.0	0.0	0.0	0.0	0.0	0.5	2.4	15.1
Lothian														
Edinburgh														
Royal Botanic Gdn.	26	5.0	5.3	1.1	0.2	0.0	0.0	0.0	0.0	0.0	0.0	0.5	1.9	14.0
Royal Observatory	134	5.5	6.1	1.8	0.3	0.0*	0.0	0.0	0.0	0.0	0.0	1.0	2.0	16.7
Penicuik	185	9.6	9.3	3.8	0.9	0.1	0.0	0.0	0.0	0.0	0.0*	1.9	4.7	30.3
Dunbar	23	1.7	2.6	0.4	0.1	0.0	0.0	0.0	0.0	0.0	0.0	0.2	0.6	5.6
Strathclyde														
Tiree	9	1.4	1.2	0.4	0.1	0.0	0.0	0.0	0.0	0.0	0.0	0.1	0.4	3.6
Glasgow Airport	5	3.5	1.8	0.7	0.1	0.0	0.0	0.0	0.0	0.0	0.0	0.6	1.3	8.0
Auchincruive (nr Ayr)	45	2.3	1.7	0.7	0.1	0.0	0.0	0.0	0.0	0.0	0.0	0.5	·0.6	5.9
Dumfries & Galloway														
Eskdalemuir	242	10.1	9.6	4.7	0.7	0.0*	0.0	0.0	0.0	0.0	0.0*	1.9	5.0	32.0

Note: 0.0* in the 30-year period indicates there was only one day with snow lying at 0900 hours in this month.

Snow covering the upper slopes of Ben Nevis

Photograph courtesy of Alex Gillespie

THE WEATH

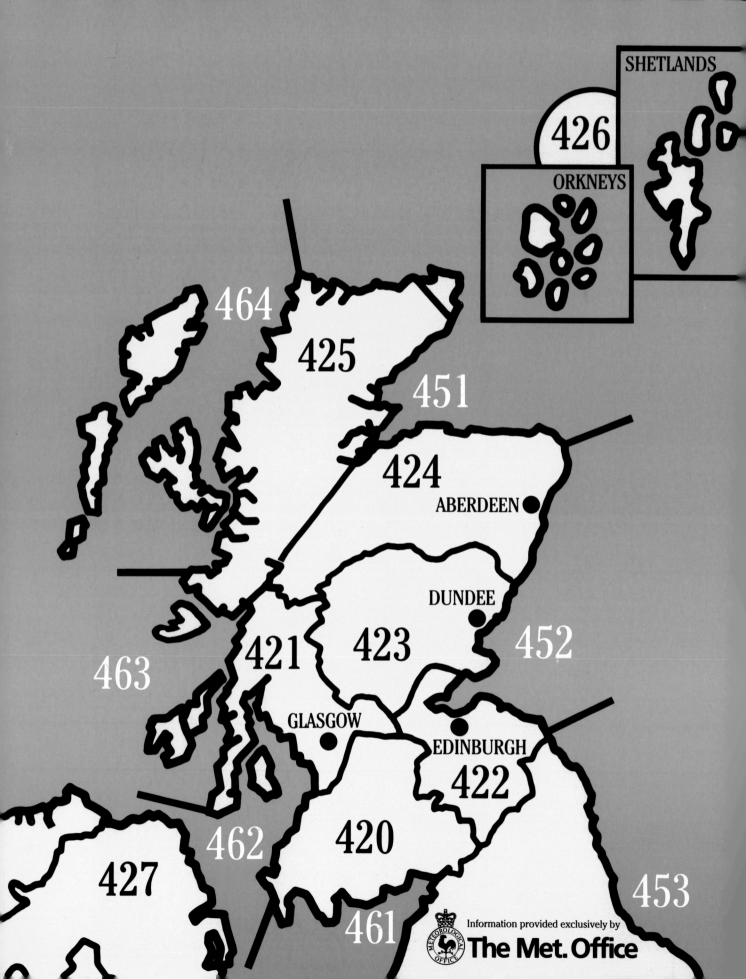

SHETLANDS

426

ORKNEYS

464

425

451

424

ABERDEEN

DUNDEE

421

423

452

463

GLASGOW

EDINBURGH

422

462

420

427

461

453

Information provided exclusively by
The Met. Office

ER, OR NOT.

Whatever outdoor activity you enjoy in Scotland, accurate weather forecasts can make a big difference to your day. All weather information for these services is supplied exclusively by The Met Office, the world leader in weather forecasting.

Weathercall provides detailed forecasts for each of the local regions shown, for up to 5 days ahead. The service is available day and night, seven days a week. Weathercall is updated several times a day to ensure you receive the latest, really accurate information.

National UK forecast 0898 500 400.
European Weathercall 0898 500 466.

Marinecall forecasts give precise information on sea state, wind strength and direction, visibility levels, high water times and the latest gale warnings. Marinecall is available at any time of day or night, seven days a week.

Skicall provides a regular update of the weather and skiing conditions in Aviemore, Glencoe, Glenshee and Lecht during the season.

FOR FREE PLASTIC CARD WITH ALL SERVICE DETAILS PHONE 01-236 3500

An important service for all those who are climbing or walking anywhere in the Highlands.
441   WEST: Lochaber, Skye, Torridon and Argyll
442   EAST: Cairngorms, East and South Highlands

Sunshine and day length

Generally, Scotland is more cloudy than England, due mainly to the much greater area of high hills, and there is no area in Scotland which can equal the sunshine durations recorded on the south coast of England. However, on the whole, the differences in sunshine duration between the two countries are quite small if stations in similar geographical locations (west coast, east coast, inland lowland, inland mountainous, etc.) are compared.

The most notable feature of the monthly averages of mean daily sunshine (**Table 3**) is the bias in favour of late spring and early summer, especially in the Western Isles and along the west coast where, on average, May is the sunniest month of the year and June is only slightly less sunny. It should also be noted that, on average, April in these areas is more sunny than the major holiday months of July and August when there is a marked fall in the mean daily sunshine from the early summer values.

The general distribution of bright sunshine in Scotland in December and June is shown in **Figures 4(a)** and **4(b)**. Although Tiree has a well-deserved reputation for high sunshine averages in May and June it can be seen that the same applies to all the western coastal fringes from the Uists in the Outer Hebrides to around the Firth of Clyde and the Solway Firth. For the year as a whole the sunniest places are near the outer estuaries of the Firths of Tay and Forth with Dunbar having the highest average annual sunshine total in Scotland.

The relatively high latitude of Scotland means that although winter days are very short, this is amply compensated by long summer days with an extended twilight. On the longest day there is no complete darkness in the north of Scotland, and sports such as tennis and golf can be played anywhere in Scotland until more than an hour later than is possible in the south of England. Lerwick in Shetland has about 4 hours more daylight (including twilight) at mid-summer than London (**Table 4**).

Figure 4(a). Daily sunshine (hours) DECEMBER

Midnight at mid-summer in Shetland

Figure 4(b).
Daily sunshine (hours)
JUNE

6.0

6.5

5.5

5.5

Less than
5.5

6.0

6.0

Less
than
5.5

5.5

6.0

6.5

Photograph courtesy of Alan Gair

Table 3. 30-year (1951–80) average duration of bright sunshine in hours for selected Scottish stations

Location	Altitude (metres)	Jan	Feb	Mar	Apr	May	Jun	Jul	Aug	Sep	Oct	Nov	Dec	Annual	Annual total
Shetland															
Lerwick	82	0.7	1.9	2.7	4.5	4.9	5.3	4.0	3.8	3.2	2.0	1.1	0.5	2.9	1056
Orkney															
Kirkwall	26	1.0	2.3	3.1	4.9	5.2	5.4	4.3	4.1	3.4	2.4	1.3	0.7	3.2	1160
Western Isles															
Stornoway	15	1.2	2.5	3.6	5.2	6.0	5.9	4.2	4.4	3.6	2.5	1.5	0.8	3.4	1256
Benbecula	5	1.3	2.5	3.7	5.8	6.5	6.5	4.6	4.9	3.8	2.5	1.6	0.9	3.7	1361
Highland															
Prabost (Skye)	67	1.3	2.7	3.5	5.2	6.0	5.8	4.1	4.3	3.3	2.3	1.5	1.0	3.4	1243
Onich (nr Fort William)	15	0.9	2.2	2.9	4.4	5.3	5.0	3.7	3.8	2.9	2.1	1.1	0.6	2.9	1059
Wick	36	1.3	2.6	3.4	5.0	5.2	5.4	4.4	4.3	3.7	2.8	1.6	1.0	3.4	1240
Nairn	8	1.4	2.8	3.6	4.9	5.3	5.6	4.7	4.3	3.9	2.9	1.8	1.2	3.5	1290
Fort Augustus	21	0.8	2.1	2.9	4.2	5.0	5.0	3.8	3.8	2.9	2.0	1.0	0.6	2.9	1044
Glenmore Lodge	341	0.8	2.4	3.4	4.4	5.1	5.2	4.2	4.1	3.5	2.6	1.1	0.5	3.1	1137
Grampian															
Braemar	339	0.8	2.0	2.9	4.5	5.2	5.5	4.9	4.2	3.4	2.1	1.1	0.6	3.1	1137
Craibstone (nr Aberdeen)	102	1.7	2.7	3.4	5.0	5.6	6.0	5.2	4.7	3.9	3.1	2.1	1.5	3.7	1367
Tayside															
Perth	23	1.4	2.3	3.2	5.1	5.7	6.0	5.5	4.5	3.7	2.7	1.8	1.1	3.6	1309
Arbroath	29	1.8	2.9	3.6	5.6	6.1	6.4	5.9	5.2	4.4	3.2	2.4	1.6	4.1	1499
Fife															
St Andrews	18	1.8	2.6	3.4	5.1	5.8	6.2	5.6	4.9	4.2	3.1	2.2	1.5	3.9	1415
Lothian															
Edinburgh (Royal Botanic Garden)	26	1.5	2.4	3.2	4.9	5.7	6.1	5.5	4.8	4.0	3.0	2.0	1.3	3.7	1351
Dunbar	23	1.8	2.8	3.7	5.4	6.1	6.7	6.1	5.4	4.4	3.4	2.4	1.7	4.2	1523
Strathclyde															
Tiree	9	1.4	2.4	3.7	5.8	6.9	6.6	5.1	5.2	3.9	2.5	1.5	0.9	3.8	1400
Glasgow Airport	5	1.3	2.2	3.1	5.1	6.0	6.2	5.3	4.7	3.7	2.6	1.6	1.0	3.6	1303
Prestwick Airport	16	1.6	2.8	3.5	5.6	6.8	6.7	5.7	5.2	4.0	2.9	1.9	1.3	4.0	1465
Dumfries and Galloway															
Dumfries	49	1.5	2.5	3.2	4.9	5.8	6.1	5.1	4.9	3.7	2.9	2.0	1.3	3.7	1338
Eskdalemuir	242	1.3	2.2	2.9	4.5	5.3	5.4	4.6	4.3	3.3	2.5	1.8	1.2	3.3	1198

Table 4. Times of sunrise (s.r.), sunset (s.s.) and end of civil twilight (c.t.)

	1 May (Times in BST)			21 June (Times in BST)			1 September (Times in BST)			21 December (Times in GMT)		
	s.r.	s.s.	c.t.	s.r.	s.s.	c.t.	s.r.	s.s.	c.t.	s.r.	s.s.	c.t.
Lerwick 60°08'N, 1°11'W	05 02	21 02	21 55	03 38	22 34	00 19	05 59	20 08	20 52	09 07	14 57	15 56
Portree (Skye) 57°28'N, 6°19'W	05 34	21 01	21 48	04 25	22 28	23 41	06 26	20 22	21 04	09 06	15 41	16 25
Dumfries 55°04'N, 3°29'W	05 34	20 49	21 32	04 34	21 57	22 56	06 20	20 08	20 34	08 37	15 41	16 33
London 51°30'N, 0°08'W	05 32	20 23	21 02	04 43	21 20	22 09	06 12	19 39	20 13	08 03	15 53	16 34

A sparkling dawn after a cold, clear night

Photograph courtesy of Mike Porter

Temperature

In winter, temperatures in the British Isles are influenced to a very large extent by the surface temperature of the surrounding seas, and since the North Sea is cooler than the waters off the west coast of Great Britain the major temperature decrease across the country is from west to east and not from north to south as might be supposed. Superimposed on this are the effects of distance from the coast (or continentality) and elevation. Normally temperature falls with height at an average rate of about 0.7 °C per 100 m for the maximum temperature and 0.5 °C per 100 m for the minimum temperature. **Figure 5(a)** shows the mean daily maximum air temperature and **Figure 5(b)** the mean daily minimum air temperature in January over Scotland for the period 1951–80. (The values have been reduced to mean sea level using the corrections given above.)

On the coldest nights however, when skies are clear, winds are light and there is a covering of snow on the ground, the lowest temperatures do not occur at the summits of the highest mountains but at the bottom of inland valleys into which dense, cold air drains. The lowest temperature reading made at a standard observing station in Britain was −27.2 °C at Braemar (339 m) in upper Deeside in February 1895 and again in January 1982. By contrast, in over 20 years of weather recording at the summit of Ben Nevis (1344 m) from 1883 to 1904 the lowest temperature was only −17.4 °C.

Full temperature statistics for a selection of stations in Scotland for the period 1951–80 are given in **Table 5**, and for comparison January and July average daily maximum and minimum temperatures for a selection of places in England, Wales and Northern Ireland are shown in **Appendix 1**. From these figures it can be seen that with the exceptions of south-west England and along the south coast, there is little difference in January temperatures between England and Scotland.

In spring, summer and autumn the effect of latitude on the heat received from the sun is the dominant factor, and temperatures in Scotland are a few degrees lower than in England, with the greatest difference occurring in summer. However, there are fewer excessively warm days and nights which may be too enervating for active outdoor activities such as golf, fishing, walking, pony trekking, or even just touring by car. Such sultry weather is quite often experienced in southern England in the summer but rarely happens in Scotland.

Figure 5(a).
Mean daily maximum
temperature (°C) January

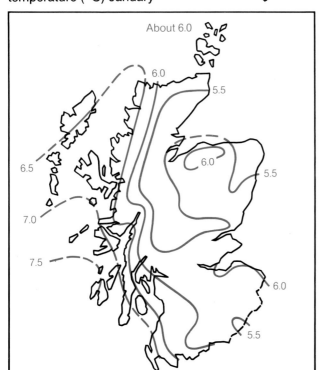

Figure 5(b).
Mean daily minimum
temperature (°C) January

Warm or even hot weather can and does occur in inland Scotland, but it is often accompanied by a very high daily range of temperature in the valleys, especially in the spring and early summer. Sometimes the temperature will fall below freezing point overnight, but rise to the mid-twenties during the day. This behaviour contrasts with the relatively small daily range in temperature on the coasts of Scotland where the moderating influence of the sea limits the fall of temperature during the night, and where sea-breezes temper the maximum temperatures on warm days during the summer. **Figures 6(a)** and **6(b)** show the mean daily maximum and minimum air temperatures during July over Scotland (reduced to mean sea level).

Figure 6(a).
Mean daily maximum
temperature (°C) July

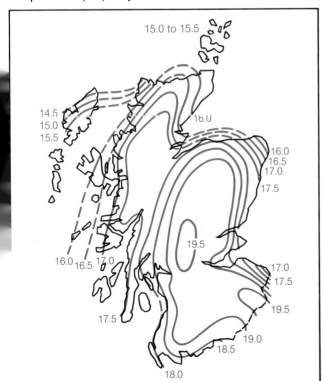

Figure 6(b).
Mean daily minimum
temperature (°C) July

Table 5. Averages and extremes of temperature (°C) for a selection of Scottish stations (1951–80)

Location	Jan	Feb	Mar	Apr	May	Jun	Jul	Aug	Sep	Oct	Nov	Dec	Year
Auchincruive													
Average Daily Maximum	6.2	6.3	8.5	11.1	14.4	17.0	17.7	17.7	15.7	13.0	8.9	7.3	12.0
Average Daily Minimum	0.9	0.7	2.2	3.7	6.1	8.9	10.5	10.5	9.0	6.9	3.3	2.0	5.4
Average Monthly Maximum	10.8	10.6	13.0	16.3	20.8	23.6	23.3	22.6	20.3	17.5	13.4	11.7	
Average Monthly Minimum	−6.4	−5.5	−3.4	−1.8	0.7	3.6	5.4	4.8	2.4	0.2	−3.2	−4.8	
Absolute Maximum	14.4	12.8	18.3	20.5	25.4	28.3	29.4	29.8	24.4	22.2	16.5	13.9	29.8
Absolute Minimum	−11.7	−10.6	−11.1	−3.9	−2.8	1.1	2.2	1.7	−1.1	−3.4	−6.7	−10.6	−11.7
Braemar													
Average Daily Maximum	3.7	3.8	6.1	9.3	12.9	16.3	17.2	16.7	14.2	10.9	6.4	4.7	10.2
Average Daily Minimum	−2.5	−3.1	−0.9	0.7	3.5	6.6	8.2	8.0	6.3	4.1	0.2	−0.9	2.5
Average Monthly Maximum	9.0	8.9	11.8	16.0	20.8	23.4	23.6	22.6	20.0	16.3	12.0	10.0	
Average Monthly Minimum	−13.3	−13.4	−9.1	−5.9	−2.3	0.5	2.2	1.0	−0.8	−3.4	−8.6	−10.4	
Absolute Maximum	11.8	13.3	17.2	21.0	24.6	27.8	30.0	30.0	23.9	23.0	15.1	12.8	30.0
Absolute Minimum	−22.2	−25.0	−21.7	−10.0	−6.7	−2.2	−0.5	−2.0	−6.0	−8.1	−14.4	−16.6	−25.0
Craibstone (near Aberdeen)													
Average Daily Maximum	5.0	5.3	7.2	9.8	12.4	15.7	17.0	16.8	14.9	12.0	7.9	6.1	10.8
Average Daily Minimum	0.2	0.0	1.4	2.6	5.1	7.9	9.4	9.5	8.1	6.1	2.6	1.2	4.5
Average Monthly Maximum	10.4	11.0	13.2	16.3	19.1	22.4	22.9	22.1	20.6	17.3	13.1	11.5	
Average Monthly Minimum	−5.8	−5.6	−3.4	−2.1	0.4	3.2	4.8	4.6	3.0	0.7	−2.4	−4.1	
Absolute Maximum	14.2	16.3	20.0	20.0	22.9	26.1	26.7	26.1	25.0	21.1	18.3	14.3	26.7
Absolute Minimum	−11.7	−12.2	−11.7	−6.7	−3.0	0.4	1.7	1.7	−0.6	−1.7	−5.0	−7.8	−12.2
Dumfries													
Average Daily Maximum	5.6	6.0	8.2	11.2	14.4	17.3	18.3	18.1	15.9	12.9	8.6	6.8	11.9
Average Daily Minimum	0.4	0.3	1.7	3.3	6.0	9.0	10.6	10.5	8.8	6.4	2.9	1.5	5.1
Average Monthly Maximum	10.5	10.3	12.9	16.4	20.9	23.9	23.9	23.0	20.4	17.3	13.2	11.5	
Average Monthly Minimum	−6.8	−5.9	−4.0	−2.1	0.5	3.9	5.5	5.3	2.6	−0.5	−4.2	−5.3	
Absolute Maximum	14.3	13.3	17.8	19.3	25.2	28.3	29.8	28.6	25.6	22.8	15.6	13.9	29.8
Absolute Minimum	−13.9	−11.1	−12.2	−3.9	−2.2	0.6	2.8	2.2	−1.1	−3.9	−9.0	−10.6	−13.9
Edinburgh (Royal Botanic Garden)													
Average Daily Maximum	6.2	6.4	8.5	11.2	14.2	17.1	18.4	18.2	16.3	13.3	9.0	7.1	12.2
Average Daily Minimum	0.4	0.3	1.9	3.6	6.2	9.3	10.9	10.7	9.0	6.5	2.6	1.5	5.2
Average Monthly Maximum	11.6	11.4	13.7	17.6	20.6	23.4	23.3	23.1	21.1	18.3	14.3	12.4	
Average Monthly Minimum	−6.6	−6.4	−4.0	−2.2	0.8	4.5	6.3	5.5	2.8	−0.3	−4.0	−5.3	
Absolute Maximum	14.4	15.0	20.0	20.9	25.2	27.5	28.3	30.0	23.9	24.4	17.3	14.4	30.0
Absolute Minimum	−12.8	−11.7	−11.1	−6.1	−1.7	1.3	4.4	2.2	−1.1	−3.7	−8.3	−11.1	−12.8
Eskdalemuir													
Average Daily Maximum	4.3	4.6	7.0	10.3	13.6	16.5	17.4	17.2	14.8	11.8	7.3	5.4	10.8
Average Daily Minimum	−1.5	−1.6	−0.1	1.4	4.0	7.0	8.7	8.7	7.0	4.7	1.0	−0.3	3.3
Average Monthly Maximum	9.1	9.4	12.6	16.5	20.8	23.6	23.4	22.6	19.9	16.8	11.8	10.2	
Average Monthly Minimum	−10.4	−9.4	−7.2	−5.6	−2.8	0.8	2.1	1.7	−0.3	−2.9	−7.1	−8.0	
Absolute Maximum	11.2	13.3	19.4	20.0	25.3	27.4	29.8	28.4	25.0	22.2	13.6	13.4	29.8
Absolute Minimum	−17.4	−16.2	−14.7	−9.8	−7.1	−2.5	−1.7	−2.2	−5.0	−7.8	−11.3	−18.5	−18.5
Glasgow Airport													
Average Daily Maximum	6.1	6.5	8.8	11.9	15.1	17.9	18.7	18.6	16.3	13.2	8.9	7.2	12.4
Average Daily Minimum	0.0	0.0	1.5	3.2	5.9	8.8	10.3	10.1	8.5	6.1	2.3	1.1	4.8
Average Monthly Maximum	11.2	11.1	13.4	17.3	21.8	24.2	24.1	23.7	20.8	17.7	13.7	12.0	
Average Monthly Minimum	−8.0	−7.1	−5.1	−3.2	−0.6	3.0	4.5	3.5	1.1	−1.7	−5.6	−6.6	
Absolute Maximum	13.3	13.9	18.9	20.2	27.4	28.9	29.2	31.2	26.7	23.9	15.6	13.3	31.2
Absolute Minimum	−17.4	−15.0	−12.5	−5.4	−3.9	1.2	0.8	1.1	−4.0	−5.2	−10.4	−12.6	−17.4
Lerwick													
Average Daily Maximum	5.1	4.9	6.0	7.8	10.0	12.5	13.7	14.0	12.5	10.3	7.4	6.0	9.2
Average Daily Minimum	1.0	0.7	1.7	2.6	5.0	7.4	9.0	9.4	8.1	6.3	3.3	1.8	4.7
Average Monthly Maximum	8.8	8.3	9.3	11.3	14.1	16.9	17.7	17.2	15.6	13.7	10.6	9.6	
Average Monthly Minimum	−4.1	−4.3	−3.2	−2.2	0.4	3.1	5.5	5.3	3.4	0.7	−2.2	−3.7	
Absolute Maximum	11.8	11.1	13.3	15.0	19.4	23.3	22.8	20.9	19.4	17.2	12.8	11.1	23.3
Absolute Minimum	−8.9	−7.3	−7.2	−5.7	−2.2	−0.6	3.5	2.9	−0.6	−2.2	−5.7	−8.2	−8.9

Location	Jan	Feb	Mar	Apr	May	Jun	Jul	Aug	Sep	Oct	Nov	Dec	Year
Nairn													
Average Daily Maximum	5.8	6.1	8.5	10.7	13.6	16.5	17.6	17.4	15.6	12.7	8.4	6.7	11.6
Average Daily Minimum	0.1	−0.4	1.4	2.8	5.6	8.4	10.1	10.0	8.3	6.1	2.4	1.1	4.7
Average Monthly Maximum	11.6	11.5	14.0	16.6	20.4	23.6	23.6	23.5	21.0	18.4	13.9	12.0	
Average Monthly Minimum	−7.5	−7.9	−4.8	−3.1	−0.1	2.5	4.9	4.2	2.2	−0.1	−3.8	−5.8	
Absolute Maximum	15.0	15.0	17.8	19.4	25.9	27.8	29.0	30.6	25.0	24.4	17.9	14.3	30.6
Absolute Minimum	−12.8	−16.7	−11.1	−6.7	−3.3	−2.2	2.0	0.5	−1.9	−3.3	−10.6	−12.2	−16.7
Onich													
Average Daily Maximum	6.3	6.7	8.6	11.2	14.5	16.6	17.2	17.3	15.4	12.8	8.8	7.3	11.9
Average Daily Minimum	0.9	0.5	2.0	3.2	5.9	8.4	9.9	9.9	8.5	6.5	3.0	1.9	5.0
Average Monthly Maximum	11.2	11.2	12.8	16.6	21.5	23.5	23.3	22.7	20.0	17.4	13.3	12.0	
Average Monthly Minimum	−5.9	−5.6	−3.8	−2.3	−0.1	3.4	4.3	4.2	2.5	−0.2	−3.6	−4.8	
Absolute Maximum	16.7	14.4	18.9	21.5	26.5	28.5	30.0	29.5	23.9	23.3	16.0	14.4	30.0
Absolute Minimum	−12.0	−10.0	−8.0	−6.5	−4.0	0.5	2.2	1.7	−2.0	−3.5	−6.7	−11.1	−12.0
Perth													
Average Daily Maximum	5.7	6.0	8.4	11.8	14.9	18.0	19.1	18.6	16.2	12.9	8.5	6.6	12.2
Average Daily MInimum	−0.6	−0.6	1.4	.3.0	5.8	8.8	10.3	10.1	8.3	5.8	1.8	0.6	4.5
Average Monthly Maximum	11.3	11.1	13.1	17.6	21.6	24.4	24.8	23.7	21.1	17.5	13.8	11.9	
Average Monthly Minimum	−8.4	−8.0	−4.6	−2.5	0.3	3.7	5.2	4.0	1.6	−1.3	−5.2	−6.6	
Absolute Maximum	14.5	15.0	18.3	21.7	27.0	27.2	30.0	28.9	26.4	22.5	16.0	14.4	30.0
Absolute Minimum	−14.9	−18.9	−13.9	−5.0	−2.8	−0.5	1.7	0.0	−3.9	−4.4	−10.0	−13.9	−18.9
Tiree													
Average Daily Maximum	7.2	7.0	8.3	10.3	12.7	14.8	15.8	16.1	14.7	12.6	9.6	8.2	11.4
Average Daily Minimum	2.8	2.5	3.5	4.6	6.9	9.3	10.8	10.9	9.9	8.3	5.1	4.0	6.5
Average Monthly Maximum	10.3	9.9	11.0	13.7	17.2	19.7	19.6	19.3	17.7	15.5	12.7	11.2	
Average Monthly Minimum	−2.2	−2.8	−1.2	−0.2	2.2	5.2	7.3	7.0	5.2	3.2	0.1	−1.2	
Absolute Maximum	12.2	11.1	13.9	17.8	22.2	23.3	26.1	23.9	21.1	18.3	14.6	12.6	26.1
Absolute Minimum	−5.9	−6.1	−4.6	−4.4	−0.3	2.5	6.0	4.9	1.7	−0.3	−3.9	−6.7	−6.7
Wick													
Average Daily Maximum	5.6	5.7	7.3	9.3	11.3	14.2	15.4	15.4	14.1	11.8	8.2	6.5	10.4
Average Daily Minimum	0.8	0.5	1.8	2.9	5.4	7.8	9.3	9.6	8.3	6.6	2.9	1.6	4.8
Average Monthly Maximum	10.0	10.1	12.1	13.8	16.7	19.6	20.2	19.9	18.8	16.5	12.5	10.8	
Average Monthly Minimum	−5.4	−5.6	−3.9	−2.6	−0.3	2.5	4.0	3.8	2.4	0.3	−3.0	−4.2	
Absolute Maximum	12.8	14.4	17.3	17.2	20.6	23.3	25.6	22.9	23.9	25.0	16.6	13.0	25.6
Absolute Minimum	−11.1	−10.6	−8.9	−7.1	−3.9	−0.4	1.7	0.5	−2.0	−2.9	−7.8	−11.7	−11.7

Temperature conversion: degrees Celsius (centigrade) to degrees Fahrenheit $°F = (°C \times \frac{9}{5}) + 32$

°C	°F	°C	°F	°C	°F
−28	−18	−8	18	12	54
−26	−15	−6	21	14	57
−24	−11	−4	25	16	61
−22	−8	−2	28	18	64
−20	−4	0	32	20	68
−18	0	2	36	22	72
−16	3	4	39	24	75
−14	7	6	43	26	79
−12	10	8	46	28	82
−10	14	10	50	30	86

19

Wind

There is a close relationship between surface isobars (lines joining points of equal air pressure) and wind speed and direction over open level terrain. However, local topography also has a very significant effect with winds tending to be aligned along well-defined valleys. For example at Edinburgh Airport winds blow most frequently from south-westerly or north-easterly directions and very rarely from north-west or south-east.

Over land, the roughness of the ground causes a decrease in the mean wind speed compared with that which occurs over the sea, with the size of the decrease depending on the nature of the terrain. In major towns and cities the overall mean speed is considerably reduced by the buildings but local funnelling may occur and the wind may gust to about the same speed as in open country. It is this gustiness which causes much of the damage to buildings and trees during major storms. In general, wind speed increases with height with the strongest winds being observed over the summits of hills and mountains.

Since many of the major Atlantic depressions pass close to or over Scotland it is not surprising that the frequency of strong winds and gales is higher than in England and Wales. However, the windiest areas are the Western Isles, the north-west coast, Orkney and Shetland which all lie at a considerable distance from the main centres of population. Even in these extreme western and northern parts of Scotland the highest frequency of gales occurs during the winter months and prolonged spells of strong winds are unusual between May and August. A 'day with gale' is defined as one on which the mean wind at the standard measuring height of 10 m above ground attains a value of 34 knots (39 miles per hour, 17.2 metres per second) or more over any period of 10 minutes during the 24 hours. **Table 6** shows the average number of days with gale during the period 1951–80 at some Scottish stations.

Not only does the strength of the wind show a strong seasonal variation, but so too does the relative frequency of winds from different directions. For example winds with an easterly component are much more frequent in May than they are in November when westerlies are strongly predominant.

Stormy seas at Esha Ness

Table 6. Monthly and annual average number of days (1951–80) with gales for selected Scottish stations

Location	Jan	Feb	Mar	Apr	May	Jun	Jul	Aug	Sep	Oct	Nov	Dec	Year
Shetland Lerwick	8.1	5.5	5.9	2.4	1.1	1.0	0.4	0.6	2.3	4.7	6.2	8.9	47.1
Highland Wick	2.1	1.5	1.5	1.1	0.2	0.2	0.1	0.3	0.9	1.3	1.7	2.3	13.2
Fife Leuchars	1.5	0.8	1.1	0.4	0.3	0.2	0.0	0.2	0.5	0.6	0.9	1.5	8.0
Strathclyde Tiree	6.9	3.7	3.4	1.4	0.5	0.3	0.3	0.5	1.6	3.3	4.7	7.2	33.8
Glasgow Airport	1.2	0.4	0.4	0.2	0.1	0.2	0.0	0.0	0.2	0.2	0.4	1.0	4.3

Extremes

Photograph courtesy of Alan Gair

Rainfall

Maximum in a day (09–09 GMT):

238.4 mm (9.39 inches) at **Sloy** Main Adit Loch Lomond (Strathclyde, 204 m) on 17 January 1974

Maximum in an hour:

90.0 mm (3.54 inches) in 55 minutes at **Eskdalemuir**

Air temperature

(Measured under standard conditions at 1.25 m above the ground)

Highest recorded:

32.8 °C (91 °F) at **Dumfries** (Dumfries and Galloway, 49 m) on 2 July 1908 (and on several occasions at other places in the 19th century)

Lowest recorded:

−27.2 °C (−17 °F) at **Braemar** (Grampian, 339 m) on 11 February 1895 and 10 January 1982

Bright sunshine

Maximum duration in a month:

329 hours at **Tiree** (Strathclyde) in May 1946 and May 1975 (An average of 10.6 hours per day)

Minimum duration in a month:

1.3 hours at **Paisley**, near Glasgow (Strathclyde) December 1890

Wind speed

Highest gust recorded at a low-level site:

123 knots (142 miles per hour, 63 metres per second) at **Fraserburgh** (Grampian, 18 m) on 13 February 1989

Highest gust recorded at a high-level site:

150 knots (173 miles per hour, 77 metres per second) at **Cairngorm** Automatic Weather Station (Grampian, 1245 m) on 20 March 1986

Appendix

Appendix 1. Climatological data for a selection of places in England, Wales and Northern Ireland (Averages over the 30-year period 1951–80)

Location	Altitude m	Average annual rainfall mm	Average daily temperature January Max °C	Min	Average daily temperature July Max °C	Min	Average annual duration of bright sunshine hours
Belfast Airport (Antrim)	68	837	6.2	0.7	18.2	10.8	1298
Douglas (Isle of Man)	85	1131	6.8	2.7	17.2	11.3	1572
Ambleside (Cumbria)	46	1865	5.9	0.2	19.1	10.8	1185
Durham (Durham)	102	645	5.5	0.1	19.1	10.3	1300
Leeming (North Yorkshire)	32	617	5.9	0.1	19.5	10.8	1331
Blackpool Airport (Lancashire)	10	847	6.3	1.0	18.7	12.1	1534
Hull (Humberside)	2	653	6.1	1.3	20.2	12.0	1380
Rhyl (Clwyd)	9	661	7.3	2.3	18.8	12.4	1475
Manchester Airport (Greater Manchester)	75	806	6.1	0.8	19.4	11.8	1359
Buxton (Derbyshire)	307	1289	4.3	−0.3	17.4	10.3	1149
Lincoln (Lincolnshire)	6	593	5.7	−0.1	20.4	10.5	1401
Skegness (Lincolnshire)	5	601	5.8	0.9	17.6	10.2	1512
Shrewsbury (Shropshire)	55	624	6.4	0.8	20.3	11.5	1349
Aberystwyth (Dyfed)	4	959	7.8	2.5	18.2	12.0	1473
Llandrindod Wells (Powys)	235	1003	5.5	0.1	19.5	10.3	1244
Stratford-upon-Avon (Warwicks)	49	627	6.3	0.1	21.3	10.7	1371
Cambridge (Cambridgeshire)	24	552	6.2	0.8	21.2	11.6	1508
Cardiff (South Glamorgan)	62	1064	6.9	1.8	20.5	12.4	1497
Oxford (Oxfordshire)	63	663	6.4	1.2	21.3	12.3	1517
London Airport (Greater London)	25	610	6.8	1.1	22.0	12.9	1494
Clacton-on-Sea (Essex)	16	542	5.7	1.5	20.1	13.2	1635
Ilfracombe (Devon)	8	1063	8.3	4.3	18.6	13.8	1631
Penzance (Cornwall)	19	1131	9.4	4.4	19.4	12.7	1738
Plymouth (Devon)	27	992	8.3	3.1	19.1	12.6	1687
Bournemouth (Dorset)	40	802	7.1	2.0	20.2	12.4	1777
Shanklin (Isle of Wight)	55	906	7.1	2.5	19.4	12.8	1908
Eastbourne (East Sussex)	7	811	7.0	2.8	19.5	13.6	1827
Folkestone (Kent)	39	727	6.5	2.0	19.9	13.3	1732